Famous in STEM

Volume 2:
Mathematical Masterminds

Javier Sanz

For all who came before, who observed the beauty of nature and dared to ask why.

To the future scientists who will dedicate their lives to the endlessly fascinating adventure of discovery so that we may understand a little more about the workings of the universe and our place within it.

To those who dare to think differently. Your radical ideas change the world.

Thank you!

Dear fellow number nerds and math curious. Remember when we thought algebra was as tough as it gets? Oh, sweet summer children we were!

If this book helps you see the beauty in equations or just makes you chuckle, we'd be irrational with joy to hear about it! Toss us a review on Amazon. Tell us which part was your favorite, or which concept you're still scratching your head over (don't worry, we are too sometimes).

☆ ☆ ☆ ☆ ☆

You can use this QR code to submit your review

TABLE OF CONTENTS

CARL FRIEDRICH GAUSS

Carl Friedrich Gauss was one of the greatest mathematicians who ever lived. Nicknamed "the prince of mathematics", his remarkable contributions revolutionized many fields including number theory, statistics, analysis, differential geometry, geodesy, magnetism, astronomy, and optics.

Gauss was born in 1777 in Brunswick, Germany. His genius was evident from a very young age. There are many fun stories that illustrate his precocious talent.

One anecdote involves Gauss in primary school. His teacher wanted to keep the bored prodigy busy so he asked the class to add up all the whole numbers from 1 to 100. Gauss came up with the solution almost instantly, while the rest of the students slowly worked through the tedious task. The teacher was astonished and asked Gauss to explain his quick thinking. Gauss described how he viewed the numbers arranged in rows, with 50 rows containing the number pairs 1 + 100, 2 + 99, 3 + 98, and so on, that summed to 101. Therefore, the total was 50 × 101 = 5050. His insightful pattern recognition shows how Gauss could come up with simple, elegant methods to solve problems.

Gauss made seminal contributions to number theory during his teenage years. At just 18 years old, he discovered that any integer can be expressed as the sum of three triangular numbers (e.g., 55 = 21 + 15 + 10). This was a major breakthrough in a 2000-year-old problem!

A triangular number is a number that, if you had that many dots, you could arrange them into a shape of an equilateral triangle:

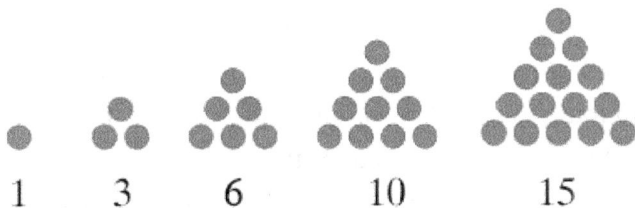

Each next triangular number adds one more row of dots. The rule for finding the nth triangular number is adding all the numbers from 1 to n. So, if you want to find the 4th triangular number, you add $1 + 2 + 3 + 4$, which equals 10.

More mathematically, triangular numbers can be defined using a formula. The nth triangular number can be found by the formula:

$T(n) = n * (n + 1) / 2$

So, for example, the first few triangular numbers are:

$T(1) = 1 * (1 + 1) / 2 = 1$
$T(2) = 2 * (2 + 1) / 2 = 3$
$T(3) = 3 * (3 + 1) / 2 = 6$
$T(4) = 4 * (4 + 1) / 2 = 10$
$T(5) = 5 * (5 + 1) / 2 = 15$

So, the sequence of triangular numbers begins 1, 3, 6, 10, 15, and so on.

In 1801 at age 24, Gauss published his magnum opus (important work of art) called *Disquisitiones Arithmeticae*, a masterpiece covering number theory and many original research contributions. It firmly established Gauss as one of the greatest mathematicians ever. He laid the groundwork for modern number theory and was so advanced that it took decades for other mathematicians to catch up, for example when he conjectured the Prime Number Theorem.

The Prime Number Theorem is a fundamental theorem in number theory. It describes how prime numbers become less frequent as they become larger by precisely quantifying the rate at which this occurs.

The Prime Number Theorem states that if $\pi(N)$ is the number of primes less than or equal to N, then as N goes to infinity, $\pi(N)$ is approximately equal to N/ln(N), where ln(N) is the natural logarithm of N. This is often written as:

$\pi(N) \sim N/\ln(N)$

A more intuitive way to understand this is that if you pick a random large number N, then the chance that N is prime is about 1 / ln(N).

Beyond number theory, Gauss made founding discoveries across many fields. In geometry, he invented differential geometry which studies curved surfaces. In physics, Gauss's law describes the relationship between electric fields and charges. He did pioneering work in geodesy, developing techniques to accurately survey the shapes of planets.

In statistics, Gauss discovered the normal distribution (also called Gaussian distribution or bell curve), enabling new methods for estimating error and testing significance. The term "normal" comes from the fact that this distribution is very common in nature and statistics. Things like heights of people, blood pressure, scores on a test, and measurement errors all follow this distribution. The mean, median, and mode are all the same and occur at the peak of the distribution.

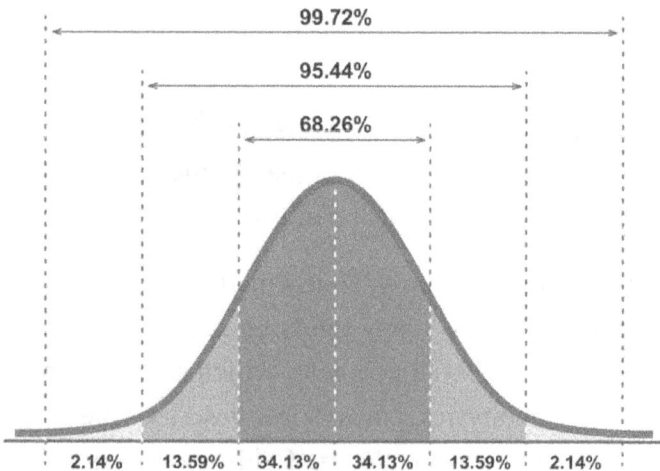

Gauss had diverse interests beyond math and science, reflecting his broad intellect. As a teenager, he learned over a dozen languages including Latin, French, German, Russian, English, Italian, Spanish, and Sanskrit. He also studied archaeology, had a deep knowledge of classical literature, and was an avid gardener.

Gauss led a fascinating personal life with some intriguing stories. He was a devoted family man who married twice and had 6 children. Always humble despite his genius, Gauss disagreed with naming mathematical concepts after people. Thus, he refused the honor of having Gaussian curve named after him, preferring just normal distribution. Amusingly, Gauss would intentionally publish only incomplete proofs to force others to make their own discoveries!

In closing, Carl Friedrich Gauss was undoubtedly one of history's greatest mathematical minds. His seminal contributions and inventions across fields like number theory, geometry, statistics, physics established the foundations for modern mathematics. Gauss's genius, creativity, and problem-solving abilities make him a legendary figure whose innovations changed mathematics forever. His intelligence combined with a deeply curious, prolific, and humble personality makes Gauss's story fascinating and inspiring.

BLAISE PASCAL

Born in 1623 in Clermont-Ferrand, France, Blaise Pascal was a mathematician, physicist, inventor, writer, and Catholic theologian. He was a child prodigy who was educated by his father, a tax collector in Rouen. Pascal's earliest work was in natural and applied sciences, where he made significant contributions to the study of fluids and clarified the concepts of pressure and vacuum.

One of Pascal's most significant mathematical contributions came in the form of the Pascal's Triangle. While the triangle itself had been previously described by Chinese mathematician Yang Hui and Persian mathematician Al-Karaji, Pascal made comprehensive applications of it in the western world. Each number in the triangle is the sum of the two directly above it, and it serves as a shorthand for several mathematical concepts, including combinations and the expansion of polynomials.

A. Ones
B. Counting numbers
C. Triangular numbers (like in page 7)

On the left-hand side, you have the powers of two. The triangle is symmetrical, and other patters that can be found include Fibonacci sequences, powers of 11 and more.

Pascal's Triangle is directly related to binomial coefficients and can be used to calculate the coefficients of the expansion of a binomial expression raised to any power. For example, if you were to expand the expression $(x+y)^4$, the coefficients of the terms in the expanded expression would be the same as the numbers in the fourth row of Pascal's Triangle. This has wide applications in probability theory, algebra, and calculus.

Another area where Pascal made significant contributions is the field of probability theory. Alongside Pierre de Fermat, Pascal helped lay the groundwork for the theory of probability, an area of mathematics that quantifies uncertainty and that today has profound importance in fields ranging from insurance and finance to medicine and computer science. Their work began as an exchange of letters discussing a problem in gambling, but the principles they developed form the basis of modern probability theory.

Pascal's work in physics was no less influential. He made significant contributions to the understanding of atmospheric pressure. His experiments demonstrated that the pressure of the atmosphere decreases with height and led to the development of Pascal's principle of pressure. This principle, fundamental in the field of hydrodynamics, states that changes in pressure at any point in an enclosed fluid are transmitted undiminished to all points in the fluid. This means that if you apply pressure to one part of the fluid, that pressure will be distributed evenly throughout the entire fluid. This principle can be observed in numerous everyday applications and technologies:

1. **Hydraulic Press:** The operation of a hydraulic press is based on Pascal's principle. If you apply a small force to a small-area piston, it creates a pressure that is transmitted undiminished across the fluid in the system to a larger-area piston. This results in a larger force at the larger piston, because force is equal to pressure times area. This principle allows the hydraulic press to amplify force, enabling it to lift or crush heavy objects.

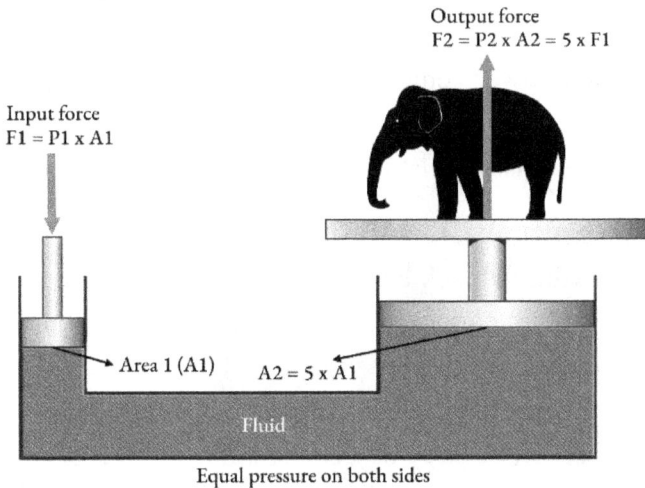

Output force
$F2 = P2 \times A2 = 5 \times F1$

Input force
$F1 = P1 \times A1$

Area 1 (A1) $A2 = 5 \times A1$

Fluid

Equal pressure on both sides

2. **Hydraulic Brakes in Vehicles:** When you press the brake pedal in your car, it pushes on a small piston that pushes brake fluid through narrow pipes to a much larger piston at each wheel. The greater area of the wheel pistons increases the force that slows and eventually stops the car.

3. **Water Towers:** A water tower operates on the same principle. The gravitational pull on the large quantity of water creates pressure at the bottom that is transmitted undiminished. This pressure is used to distribute water to houses in a neighborhood even if they're a good distance away from the tower.

4. **Hydraulic Lifts:** Hydraulic lifts, like those used in car repair shops, use Pascal's principle to raise and lower vehicles. A small force applied to a small-area piston leads to a large force at a large-area piston, just like in the hydraulic press.

5. **Hydraulic Jacks:** Like the hydraulic lift, a hydraulic jack lets you lift a heavy object by applying a modest force over a longer distance. It trades force (what you put in) for distance (what you get out), so a small amount of force, or work, can lift a heavy object a small distance.

In all these examples, the key concept is the transmission and amplification of pressure in a fluid, which is the core of Pascal's principle.

$$P = \frac{F}{A}$$

Where P = Pressure; F = Force; A = Area

Outside of his scientific and mathematical work, Pascal also explored philosophy and theology. His thoughts on religion and philosophy, especially his wager argument for belief in God, are recorded in his unfinished notes posthumously published as the "Pensées" or "Thoughts."

Pascal died in 1662 at just 39 years old, his health having deteriorated after his intense religious phase. Despite his short life, Pascal pioneered key mathematical and scientific concepts that deeply influenced Enlightenment thinkers like Leibniz and Newton. For his foundational contributions to calculus, probability, physics, and religious philosophy, Pascal remains one of the most versatile intellects of 17th century Europe.

ALAN TURING

Alan Turing was born in London on June 23, 1912. Like the workings of a complex machine or a captivating cipher, there was always more than what met the eye with Turing. His story intertwines brilliance and secrecy, success, and tragedy, creating a tale that is as fascinating as the man himself.

As a child, he was curious and eccentric child. He had a deep fascination with science and numbers from an early age. Legend has it that when a general strike in Britain prevented him from going to school, a 14-year-old Turing cycled 60 miles to school, determined not to miss a day!

Turing studied mathematics at King's College, Cambridge, where he began his path to becoming one of the most influential mathematicians and computer scientists of the 20th century. In 1936, he created the concept of a "universal machine" that could compute anything that is computable. This invention is considered the theoretical basis for the modern computer.

Turing was a marathon runner! He was very close to making the British team for the 1948 Olympics before an injury thwarted his chances. He used to amaze his colleagues by running 40 miles to London when he was needed for high-level meetings!

During World War II, Turing worked as a codebreaker at Bletchley Park, Britain's top-secret code-breaking center. He played a vital role in deciphering the messages encrypted by the German machine "Enigma," which was considered unbreakable.

Figure 1. Enigma machine

The Enigma machine was a device used for encryption and decryption, most famously by Germany during World War II. Its purpose was to protect sensitive military, diplomatic, and other communications from being read by the enemy.

The Enigma machine looked a bit like a typewriter, with a keyboard for inputting plaintext messages and a board with lights to display the encrypted output. Inside, it contained a set of rotating mechanical rotors, which scrambled the input in a complex and changeable way.

The key to the Enigma machine's security was that the settings of the rotors, and sometimes other components, could be changed regularly, according to a secret schedule known only to the operators. This meant that even if someone else had an Enigma machine, they couldn't decrypt the messages without knowing the correct settings.

How did Turing crack the Enigma? He built an electromechanical machine, known as the "Bombe," which significantly reduced the work of codebreakers. It's estimated that Turing's work shortened the war by more than two years, saving countless lives. Even Winston Churchill praised Turing's work, saying it was critical to the Allied victory. However, much of Turing's contributions remained classified for years after the war.

Figure 2. Codebreaker deciphering messages at Bletchley park

While this is not a "mathematical" achievement in the traditional sense, it required extensive knowledge of mathematics, logic, and early computer science.

Post-war, Turing turned his mind towards the future. He dreamt of machines that could not only compute but also imitate human intelligence. This led him to propose the famous "Turing Test" for machine intelligence in 1950.

The Turing Test is used to determine whether a machine can demonstrate intelligent behavior equivalent to, or indistinguishable from, that of a human.

The traditional version of the test involves three participants: a human evaluator, a human participant, and a machine. Both the human participant and the machine try to convince the evaluator that they are the human. The conversations are typically text-based to avoid bias from voice or appearance. If the evaluator cannot reliably distinguish which participant is the machine, then the machine is said to have passed the test.

It's important to note that passing the Turing Test does not necessarily imply true understanding or consciousness on the part of the machine. It simply means that the machine's responses were indistinguishable from those of a human in that specific context. It's a test of apparent intelligence, not actual intelligence.

Turing's work was the foundation of artificial intelligence, and the Turing Test is key today in discussions about AI.

His life took a tragic turn in 1952 when he was prosecuted for being homosexual, which was still criminalized in the UK. He was convicted and chose chemical castration over imprisonment. This was a profound injustice not just for any person, but even more so for a man who had given so much to his country and the world.

Two years later and at just 41 years old, Turing died by biting an apple laced with cyanide. Given an apple was found half-eaten next to his bed, it is believed his death was intentional and related to his unfair conviction. This is the bitten apple that many people believe inspired Apple's logo.

Turing's life and legacy are a testament to the indomitable spirit of human intellect and courage. While his life was marked by hardship, his work has had a lasting impact, and he's celebrated today as a pioneer of modern computing and a war hero.

In 2013, Turing was posthumously pardoned by Queen Elizabeth II, and in 2020, he was chosen to appear on the new design of the Bank of England's £50 note.

Today, Turing is considered the father of theoretical computer science and artificial intelligence. The Turing Award, computer science's version of the Nobel Prize, was named after him.

Turing's life inspires two lessons:

1. One person really can make a huge difference in the world. His work saved countless lives and shaped the computer age.

2. Prejudice against those who are "different" can cost society the gifts and talents of its own people. Turing helped save his country, only to be cruelly punished later for simply being homosexual.

Through the lens of Alan Turing's life, we see a man who changed the world through his intellect and resilience. Remember, whether you're sending a text from your smartphone or playing a video game, you're using technology that, in some way, was pioneered by the extraordinary genius of Alan Turing.

LEONHARD EULER

Euler, born on the 15th of April 1707, in Basel, Switzerland, was the firstborn in a humble pastor's family. A prodigious talent emerged in this child, discovered by his father, a man well-versed in mathematics himself, but nowhere close to the intellectual depth that his son would reach.

Euler's intellectual voyage truly began when his family moved to Riehen, where he attended school. Here, a young, spark-eyed Euler was introduced to Johann Bernoulli, the then leading light in calculus. Bernoulli, sensing Euler's innate aptitude for mathematics, took him under his wing. The foundations of Euler's brilliant career were laid in these formative years, with Bernoulli's sage advice and Euler's insatiable thirst for knowledge leading the way.

Euler's first significant accomplishment came in 1727, when he was invited to St. Petersburg to join the Academy of Sciences. It was a period of turmoil, where political strife gripped the city. However, amidst this chaos, Euler found his oasis of tranquility in the vast landscapes of mathematics. Here, he made significant contributions to infinitesimal calculus, setting the stage for the mathematical miracles to follow.

In St. Petersburg, Euler's life was as tempestuous as the equations he tamed. He married Katharina Gsell in 1734, and together they had thirteen children. Tragically, only five survived infancy, a testament to the harsh realities of 18th-century life. Euler's family was his haven, their shared laughter and joy becoming the melody that accompanied his intellectual symphony.

Euler's time in St. Petersburg was briefly interrupted when he was invited to Berlin by the Prussian monarch, Frederick the Great, in 1741. The Prussian king aimed to make Berlin the "Athens of the North," a hub of knowledge and intellectual discourse. Despite the king's notorious distaste for mathematics, he couldn't help but recognize Euler's genius.

Among Euler's achievements in Berlin, one particular work stands out - the publication of his "*Introductio in analysin infinitorum*." This book transformed the face of mathematics, providing it a much-needed coherent structure. In this masterpiece, Euler boldly defined a function, a term we take for granted today, introducing an entirely new way of understanding mathematical relationships.

An interesting anecdote from this period involves a famous dispute between Euler and the French encyclopedist Denis Diderot. Diderot, a nonbeliever, was visiting the Russian court when a debate ensued over religion's place in society. As legend has it, Euler, who was deeply religious, stepped forward and proposed a complex mathematical formula, proclaiming it as proof of God's existence. Diderot, unable to decipher it, was left speechless, leading to a humorous victory for Euler.

In 1766, Euler returned to St. Petersburg, where tragedy struck - he lost sight in his right eye. Ironically, the man who saw through the complexities of mathematical equations was increasingly trapped in a world of darkness. However, with characteristic determination, Euler never let this setback interfere with his work. In fact, some of his most profound ideas emerged during this period. Blind but undeterred, Euler's intellect illuminated the darkness around him.

Euler's monumental contributions to mathematics and physics are too numerous to recount. He developed the concept of a mathematical function, solved the Seven Bridges of Königsberg problem giving birth to graph theory, formulated Euler's Identity, recognized as the most beautiful equation in mathematics, and made significant contributions to optics, acoustics, and celestial mechanics. His legacy is etched not only in textbooks but in the minds of every student who embarks on the journey to understand the universe's numerical rhythms.

Euler's Identity: Often described as the most beautiful equation in mathematics, Euler's Identity links five fundamental mathematical constants: 0, 1, π, e, and i (the imaginary unit), in the simple equation $e^{i\pi} + 1 = 0$. It was a pioneering amalgamation of algebra, geometry, and analysis and is a testament to Euler's ability to connect seemingly disparate mathematical fields.

Euler's Formula for Polyhedra (Euler's Theorem): This simple yet profound formula, "Vertices - Edges + Faces = 2," links the number of vertices, edges, and faces of a polyhedron. This groundbreaking formula laid the foundation for the development of topology, a field of mathematics concerned with the properties of space that are preserved under continuous transformations. For example, a tetrahedron has 4 vertices, 6 edges, and 4 faces. Therefore, 4 - 6 + 4 = 2.

Euler's Number (e): Euler did not discover the number 'e' but popularized it as the base of natural logarithms. Its approximate value is 2.71828, and it serves as the cornerstone of calculus. Euler's number is used in many mathematical models, from calculating compound interest to modeling population growth.

Another one of Euler's most critical contributions to calculus was the introduction of mathematical notation. Sound simple? It was far from it. Euler established notation that brought coherence and ease into complex mathematical expressions, forever changing how mathematicians communicated. For example, he introduced the use of 'e' for the base of natural logarithms, 'i' for the square root of -1, and the Greek letter Σ for summation. His introduction of the symbol π in trigonometry is possibly his most recognizable contribution.

On September 18th, 1783, at the age of 76, Leonhard Euler breathed his last while playing with his grandchildren. His final words, "I die," were as simple and profound as the man himself. His life was a testament to the power of the human intellect and the beauty of mathematical exploration.

In Euler's life and work, we witness a relentless pursuit of knowledge and a deep fascination for the mysteries of the universe. In many ways, his life was a mathematical equation, filled with constants of familial love, variables of political change, and the unknowns of death and loss. Each part of his journey contributes to the whole, much like the integers in his beloved equations.

Euler's life can be seen as an equation yet unsolved. There are gaps in our understanding, pieces of his puzzle still missing. For even as we delve into Euler's extraordinary life and accomplishments, we are left wondering - what secrets did this grandmaster of mathematics take to his grave? Only Euler's spectral shadow, cast over the centuries, holds the answer.

PAUL ERDŐS

Paul Erdős, born in 1913 in Budapest, Hungary, is a legend in the field of mathematics. His prodigious outputs, eccentric lifestyle, and unique collaborative approach to mathematics have made him a unique figure within the mathematical world.

Erdős was a child prodigy, showing a keen interest in math from a very young age. By four, he could calculate how many seconds a person had lived if given their age. He pursued mathematics with singular determination, earning his Ph.D in that subject by the age of 21 and from the University of Budapest.

Erdős was exceptionally prolific, having written or co-written 1,525 scholarly papers during his career – a record that still stands. Even in his 70's, he would publish more papers in a year than most mathematicians do in their entire lives. This remarkable output was enabled by his itinerant lifestyle. Erdős had no permanent residence nor family of his own. Instead, he spent most of his life traveling between scientific conferences, universities, and the homes of fellow mathematicians all over the world, collaborating on work and living out of a suitcase. He would just show up at someone's house in the middle of the night ready to solve problems.

Unlike many mathematicians who concentrate on one area of mathematics, Erdős contributed to various fields, such as number theory, combinatorics, probability, and set theory. His most famous work is probably in discrete mathematics, particularly graph theory, where he introduced many fundamental concepts.

One of the first and most famous problems in graph theory is the Seven Bridges of Königsberg. The city of Königsberg in Prussia (now Kaliningrad, Russia) was set on both sides of a river, and included two large islands connected to each other and the mainland by seven bridges. The problem was figuring out if it was possible to take a walk through the city of Königsberg crossing each of its seven bridges only once. How would you do it?

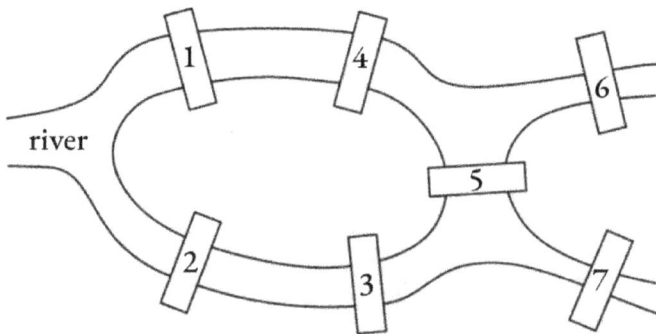

The Swiss mathematician Leonhard Euler was able to prove that such a walk is not possible, laying the foundations of graph theory. Euler realized that the route inside each landmass doesn't matter, only the sequence of bridges crossed. He represented each landmass as a vertex, and each bridge as an edge. This transformed the problem into one of finding a walk through the graph that touches every edge once. This kind of route is now known as an "Eulerian path".

Graph theory is also crucial in understanding network structures, like social networks or the internet, and flows in networks, like transportation networks or electrical grids. It's central to modern coding theory, which is fundamental to the design of the digital communications and data storage systems we use every day.

Another one of the most enduring parts of Erdős's legacy is the Erdős number, which describes the "collaborative distance" between a person and Erdős, based on joint authorship of mathematical papers. Erdős himself has an Erdős number of 0. If you co-authored a paper with Erdős, your Erdős number is 1. If you didn't work with Erdős but did write a paper with someone who had an Erdős number of 1, your Erdős number is 2, and so on. This fun concept emphasizes Erdős's extensive collaboration throughout his career. Bill Gates' and Stephen Hawking's Erdős number is 4.

Erdős was known for his eccentricities. He worked up to 19 hours a day and survived largely on caffeine, taking stimulants to keep himself awake. For him, a mathematician was a machine that turns coffee into mathematical theorems. He was also known for his humor and created his own vocabulary. For instance, he referred to children as "epsilons," the term used for a small quantity in mathematics.

Paul Erdős passed away in 1996 at age 83, leaving behind an immense and influential body of work. Paraphrasing Albert Einstein's famous quote[1], Erdős said "God may not play dice with the universe, but something strange is going on with the prime numbers".

His life, dedicated to the pursuit of mathematical truth and the building of a global mathematical community, continues to inspire mathematicians worldwide.

1. *Albert Einstein's said "God does not play dice with the universe" as a reaction to Neils Bohr's probabilistic interpretation of quantum mechanics. Find more about physicists Albert Einstein and Neils Bohr on* **Famous in STEM: the Greatest Physicists***.*

ERATOSTHENES

Eratosthenes was a prolific polymath and pioneer of ancient Greece who excelled in many fields including mathematics, astronomy, geography, philosophy, and poetry. Nicknamed "Beta" because he was renowned as the second best scholar in Greece, Eratosthenes made groundbreaking discoveries about the size of the Earth and distant stars.

Eratosthenes was born in 276 BC in Cyrene, a Greek colony located in modern-day Libya. He studied philosophy in Athens under the tutelage of Ariston of Chios before becoming the chief librarian at the Great Library of Alexandria. This important scholarly role gave Eratosthenes access to a wealth of knowledge and books from across the ancient world.

One of Eratosthenes' major accomplishments was accurately measuring the circumference of the Earth. Previous estimates of the Earth's size varied wildly, from around 18,000 miles to over 150,000 miles. Eratosthenes devised an ingenious method using basic geometry. He knew that on the summer solstice at noon in the city of Syene (modern Aswan, Egypt), the Sun would be directly overhead, casting no shadows. Meanwhile in Alexandria, shadows showed the Sun was slightly off vertical. Eratosthenes hired surveyors to measure the Sun's angle and determine the distance between the cities. Trigonometry revealed the Earth's circumference to be roughly 25,000 miles - less than 1% error from today's measurements! This pioneering achievement in geography was remarkable for its time.

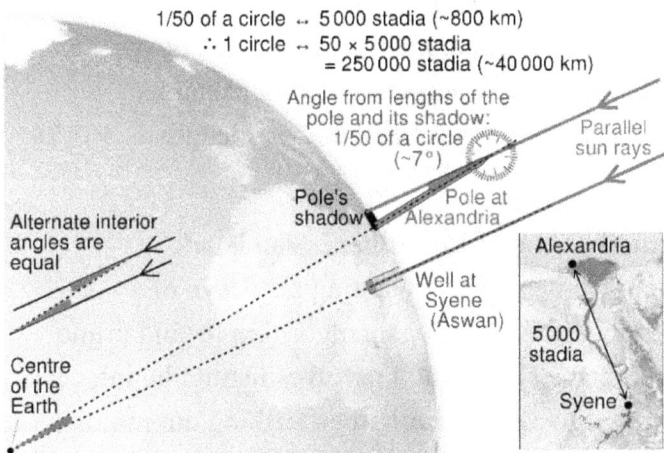

1/50 of a circle ↔ 5000 stadia (~800 km)
∴ 1 circle ↔ 50 × 5000 stadia
= 250000 stadia (~40000 km)

Angle from lengths of the pole and its shadow:
1/50 of a circle
(~7°)

Parallel sun rays

Pole's shadow

Pole at Alexandria

Alternate interior angles are equal

Well at Syene (Aswan)

Alexandria

5000 stadia

Centre of the Earth

Syene

Figure 3. Illustration of the method Eratosthenes used to calculate the circumference of the Earth by CMG Lee. Wikimedia

In astronomy, Eratosthenes calculated the distances to the Sun and Moon. He observed that at noon on the solstice in Syene, sunlight shone down a deep well, showing the Sun was directly overhead. By measuring the Sun's angle of elevation in Alexandria, he used similar triangles to estimate the distance to the Sun as 804 million stadia. This was the first serious attempt to scientifically gauge astronomical distances.

Eratosthenes also determined the tilt of the Earth's axis with impressive accuracy and proposed a leap day system to reconcile the mismatch between lunar and solar calendars. He created maps of stars and constellations, cataloging over 800 stars. Eratosthenes determined the year length as 365.24 days.

Another one of Eratosthenes' landmark achievements was inventing the Sieve of Eratosthenes, an elegant algorithm to find prime numbers. The method involves listing the integers from 2 to a given limit, then striking out multiples of each prime number in turn. The remaining unmarked numbers are prime. This simple, iterative approach was likely the first algorithmically-defined method to enumerate primes. The Sieve of

Eratosthenes proved tremendously influential and is still used today due to its efficiency and insights into number theory.

Eratosthenes wrote books compiling knowledge on mathematics and astronomy. He studied perfect numbers and amicable numbers.

A perfect number is a positive integer that is equal to the sum of its proper divisors, excluding the number itself. The smallest perfect number is 6, because its proper divisors (1, 2, 3) add up to 6. The next perfect number is 28 (1+2+4+7+14=28). Perfect numbers are rare, and as the numbers get larger, perfect numbers become increasingly scarce.

Amicable numbers are a pair of numbers where each number is the sum of the proper divisors of the other. For example, 220 and 284 are an amicable pair. The proper divisors of 220 are 1, 2, 4, 5, 10, 11, 20, 22, 44, 55, and 110, which add up to 284. The proper divisors of 284 are 1, 2, 4, 71, and 142, which add up to 220. So, each number is the sum of the proper divisors of the other, fitting the definition of amicable numbers. Like perfect numbers, amicable pairs are also rare.

Despite going blind in old age, Eratosthenes continued his work by devising geometric proofs using tactile diagrams. Some fun anecdotes describe how he loved eating meals with friends, but refused to eat if the group numbered an odd total!

Eratosthenes made many contributions to mathematics and astronomy through inspired observation, logical reasoning and an interdisciplinary approach. He embraced hands-on measurement, estimation and calculation to pioneer new methods. Eratosthenes combined ingenuity with a systematic, experimental spirit that was ahead of his time.

PYTHAGORAS

Everyone knows about Pythagoras' theorem from geometry class, right? But how much do we know about the man behind this groundbreaking concept? We're about to dive into the life of Pythagoras of Samos, the intriguing ancient Greek mathematician and philosopher.

Born around 570 BC on an island called Samos, off the coast of modern-day Turkey. Pythagoras later moved to the city of Croton in southern Italy where he established a mystical and philosophical school and religious sect known as the Pythagoreans. The Pythagoreans lived by strict rules and worshipped numbers, believing them to be the basis of everything.

Pythagoras believed that numbers were the essence of everything and that mathematical relationships could be used to understand the cosmos. The Pythagoreans summed this up in the phrase "All is number."

Now, about the theorem that made him a star of high school math textbooks – the **Pythagorean theorem**. This is the most famous of Pythagoras's achievements. The theorem states that in a right-angled triangle, the square of the hypotenuse (the side opposite the right angle) is equal to the sum of the squares of the other two sides (often described as $a^2 + b^2 = c^2$). This theorem is a fundamental principle in Euclidean geometry and has laid the groundwork for many theories and applications, from trigonometry to physics to engineering and even video game design!

But here's the kicker – Pythagoras himself might not have come up with the theorem. There's evidence that the Babylonians knew about it centuries before him. Pythagoras, however, was the one who provided the first recorded proof, hence the name.

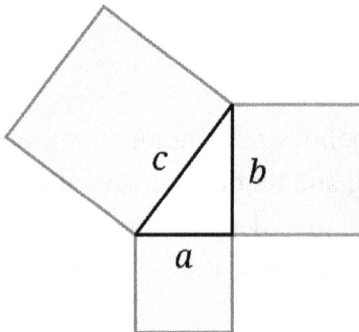

The Pythagoreans also discovered the existence of irrational numbers, i.e., numbers that cannot be expressed as a ratio of two integers and have an infinite number of non-repeating decimals. The story goes that Hippasus, a Pythagorean, discovered this when trying to express the square root of 2 as a fraction. This challenged the Pythagorean belief that all things could be reduced to whole numbers and their ratios, causing a fundamental crisis in their school of thought. Examples of irrational numbers are: pi ($\pi = 3.14159265...$), $\sqrt{2}$, $\sqrt{3}$, Euler's number ($e = 2.718281...$).

Music was an important part of Pythagorean teachings. Pythagoras is credited with discovering the relationship between the length of a vibrating string and the pitch of the note it produces. He noted that when the lengths of vibrating strings are ratios of whole numbers, the tones produced are harmonious. This formed the basis of the Western musical scale.

In the category of interesting stories, we can't skip the famous Pythagorean cup. This cup was supposedly designed by Pythagoras to promote moderation. If filled only up to a certain point, the drinker could enjoy their beverage peacefully. However, if they got greedy and filled it too much, the cup would empty its contents from the bottom!

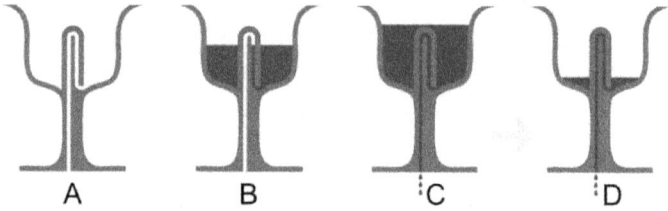

Figure 4. Cross section of a Pythagorean cup being filled: at B, it is possible to drink all the liquid in the cup; but at C, the siphon effect causes the cup to drain. By Nevit Dilmen. Wikimedia.

An odd fun fact is that beans were forbidden in the Pythagorean diet. His followers were prohibited from eating, touching, or even walking through bean fields. Why? There's no definitive answer, but one theory is that they banned beans because they look vaguely like testicles.

The mysterious mathematician-philosopher known as Pythagoras was much more than a name in a math textbook. He was a thinker, a leader, and a man of unshakable principles.

Math isn't just about cold hard facts and calculations. It's about people with their quirks, fears, and beliefs. People like Pythagoras. So, the next time you're stuck solving a triangle in geometry class, take a moment to ponder the vibrant and curious mind of Pythagoras, a man who lived over 2,500 years ago! His story might just make you see math – and beans – in a new light!

44

FIBONACCI

Going into the middle ages, we stumble upon the name Leonardo of Pisa, more commonly known by his nickname Fibonacci, which loosely translates to "son of Bonacci". Born around 1170 in Pisa, Italy, Fibonacci played an instrumental role in bringing Indian-Arabic numerals to the western world. His journey through the world of mathematics was fascinating and impactful, and his story is rich with intrigue and innovation.

Leonardo was the son of a Pisan merchant who served as a customs officer in North Africa. His early education took place in North Africa, under the guidance of Arab mathematicians. It was here that he learned about the Hindu-Arabic numeral system, which intrigued him much more than the traditional Roman numerals used in Europe at the time.

In his book "Liber Abaci," published in 1202, Fibonacci introduced Europe to this new system of numeration. The book comprehensively discussed the benefits of using the base-10 system, decimal fractions, and place value. It was through this book that Europeans were introduced to '0' and the numbers '1' through '9' as we write them today. This may not sound like a big deal now, but imagine trying to do complex arithmetic with Roman numerals!

"Liber Abaci" also contains a broad range of mathematical and geometric instructions, including a section on how to calculate square roots, practical advice on bookkeeping for businessmen, and several problems related to proportions. It demonstrated the immense practical benefits of the Hindu-Arabic numeral system, like simplifying business calculations and conversions of measurements and currencies. This work played a crucial role in replacing the use of Roman numerals, marking a significant shift in how Europeans approached mathematics and commerce.

But Fibonacci's most famous contribution to mathematics comes in the form of a simple sequence of numbers named after him - the **Fibonacci sequence**. In this sequence, each number is the sum of the two preceding ones. It starts like this: **0, 1, 1, 2, 3, 5, 8, 13, 21**, and so on. This sequence was introduced in a seemingly modest problem about rabbit population growth in "Liber Abaci," but its significance is anything but modest.

The Fibonacci sequence has deep connections with the 'Golden Ratio,' a mathematical ratio often found in nature, art, and architecture. The ratio of two successive Fibonacci numbers approximates the Golden Ratio, which is approximately **1.618**. This mathematical concept has found its way into various fields:

1. **Nature's Design**: Observe a sunflower's seed arrangement, the whorl of a seashell, a pineapple's surface, or a pinecone's bracts. You'll often find the number of spirals corresponding to consecutive Fibonacci numbers. This pattern ensures optimal packing and exposure to sunlight.

2. **Art and Architecture**: The Golden Ratio, intimately linked with the Fibonacci Sequence, has been a guide for aesthetic proportion in art and architecture. From the Parthenon to Salvador Dalí's masterpieces or the The Great Pyramid of Giza. All of them were created using the golden ratio. You can also find it on National Geographic's logo, where the ratio between the width and the height of the logo's inner rectangle is 1.62, as you can see in the following image:

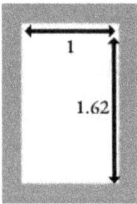

3. **Financial Markets**: Traders often use Fibonacci retracement levels as a tool to predict trends in the stock market. By applying the ratios derived from the Fibonacci Sequence, they identify potential support and resistance levels.

4. **Music**: The Fibonacci Sequence has even danced its way into musical compositions.

Some composers have structured their works around this pattern, finding in it a natural rhythm and harmony.

5. **Computer Algorithms**: In computer science, the Fibonacci numbers are used in algorithms to optimize data structures and sorting techniques, reflecting efficiency in both natural and artificial systems.

6. **Human Anatomy**: Some have noted that the proportions of various human body parts, like the lengths of our fingers' segments, often align with the Golden Ratio, an echo of the Fibonacci Sequence.

7. **Weather Patterns**: Surprisingly, the Fibonacci Spiral can be found in the formation of hurricanes and even galaxies. The way they spiral adheres to the same mathematical principle.

8. **Fractals and Chaos Theory**: The sequence's recursive nature lends itself to the creation of fractals and exploration within chaos theory, weaving complexity from simplicity.

9. **Literature and Entertainment**: From novels to movies, the Fibonacci Sequence

often appears as a plot device or thematic element, adding a layer of complexity and intrigue. In Dan Brown's novel *The Da Vinci Code* (2006), the numbers are used to unlock a safe.

10. **Optics**: In some specialized optical systems, Fibonacci numerical arrangements are used to control the distribution of light efficiently.

Fibonacci made several other key contributions to medieval mathematics:

- He promoted the usage of plus (+) and minus (-) symbols to represent positive and negative magnitudes, anticipating modern algebraic notation.

- His free promotion of the Indian number system was crucial in challenging resistance from abacists who insisted on using the abacus for calculation.

- His books demonstrated practical applications like checking the authenticity of coins, profit margin calculations, and problems involving perfect numbers.

- Fibonacci numbers proved the utility of the new Indo-Arabic mathematics in natural science applications like modeling rabbit breeding.

- Through studying inaccessible problem texts from Greece, North Africa and the Middle East, Fibonacci introduced more advanced mathematical concepts to medieval Europe.

- His work was a bridge between the ancient Greek traditions of geometry and the new Arabic algebraic traditions exemplified by figures like Al-Khwarizmi.

Fibonacci's death, thought to have occurred sometime after 1240, marked the end of a life dedicated to numerical exploration. However, his legacy continues to resonate, from the classroom to the cosmos.

In many ways, Fibonacci served as a focal point for the transmission of mathematical ideas between the Islamic world and Europe in the Middle Ages. Through his influential books and service at Frederick II's court, he substantially accelerated the adoption of Arabic numerals, arithmetic operations, algebra and number theory across Europe.

The story of Fibonacci is a sequence of fascination, discovery, perseverance, and legacy. It weaves through the realms of mathematics, nature, and history, connecting dots that form a beautiful pattern, much like the sequence he is known for.

Yet, as we reflect on Fibonacci's life and contributions, there remains a lingering enigma. A man who revealed such profound numerical beauty left behind a life hidden in mystery. Little is known about his personal life, his relationships, or the deeper thoughts that fueled his mathematical pursuits. Even his final resting place is unknown. The Fibonacci Sequence spirals out into infinity, and so too does the intrigue around the man who discovered it. What hidden patterns and undiscovered insights still lie within the depths of Fibonacci's mathematical world? That, dear reader, is a sequence yet to be unraveled.

EUCLID

Now going even further back in time to the flourishing intellectual hub of Alexandria in ancient Egypt, and there we find Euclid, a figure often hailed as the "Father of Geometry". Euclid's life and work, much like the geometrical theorems he penned, stand as timeless contributions to the understanding of the space we inhabit.

Euclid, born around 300 BCE, lived in an era when Greek culture and knowledge were spreading throughout the Mediterranean world. Although details about his personal life remain hazy, his intellectual legacy is solidified in a series of books that have deeply influenced mathematical thought for more than two millennia.

Euclid taught mathematics in Alexandria and opened his own school of geometry. He would train some of the period's brightest mathematical minds like Archimedes who then built upon his core ideas. Euclid came to be regarded as the leading mathematician in Alexandria.

Euclid wrote the most successful textbook in history. For over 2000 years, the Elements provided the backbone of geometric education and contributed key foundations for the development of mathematics, physics and astronomy. The Elements a 13-volume textbook covering the entirety of ancient Greek mathematics. The Elements begins with definitions of key terms and geometric axioms. This is followed by propositions proving major mathematical theorems using logic and rigorous proof.

The first 6 books focus on plane geometry, including important results like the Pythagorean theorem, properties of squares, triangles and circles, and constructing figures using compasses and straightedges. The next 3 books cover number theory and include Euclid's elegant proof that there are infinitely many prime numbers. The final 4 books deal with more advanced topics like solid geometry and perspective.

The Elements stood out for its systematic deductive approach of starting with self-evident axioms and building up complex theorems through irrefutable logical steps. Each proposition was accompanied by a mathematical proof. This became the standard for rigorous mathematical and scientific thinking. The Elements' organization of all known geometric knowledge also set a precedent for future encyclopedic scientific works like Newton's *Philosophiae Naturalis Principia Mathematica* (find more about Newton's work on 'Famous in STEM: The Greatest Physicists').

The Elements was translated into Latin then Arabic and was required reading for European mathematicians, philosophers, theologians, artists, and anybody seeking a comprehensive education.

Euclid's Elements and his extant works had profound historical impacts:

- They transmitted the mathematics of ancient Greek mathematicians to future generations. Euclid unified this knowledge into a canon still studied today.

- The logical, deductive approach became the exemplar for rational thought applied to philosophy, theology and the sciences for 2000 years.

- They were the main source of geometric education until the 20th century, shaping great thinkers like Galileo, Copernicus, Kepler, Newton and Einstein.

- Abraham Lincoln taught himself geometry by reading Euclid to sharpen his reasoning skills, which influenced his political oratory.

- They had a strong influence on infinity and the development of calculus by mathematicians like Newton and Leibniz.

- Artists and architects used Euclidean geometry as an underpinning for techniques of composition, perspective and design.

- They helped preserve mathematical knowledge through the Middle Ages until the Renaissance by being copied in Byzantine and Islamic manuscripts.

While little is known of Euclid the man, the Elements cemented Euclid's legacy as a pivotal figure in the history of mathematics. By giving future generations both the theorems and the tools of deduction, Euclid enabled much of the later progress in geometry, physics, philosophy and intellectual thought. For establishing a gold standard of logical reasoning and systematic scientific exposition, Euclid deserves recognition as one of history's most influential mathematicians.

GIROLAMO CARDANO

Turn the pages of history to the vibrant and transformative period of the Renaissance, and you will find an intriguing figure who mastered many disciplines: Girolamo Cardano. He was a polymath who made fundamental contributions to algebra, probability theory, and number theory, while also being a pioneering mind in disciplines like medicine and natural philosophy.

Cardano was born in Pavia, Italy in 1501. His mother was a widow and his father was a mathematically gifted lawyer who gave Girolamo an excellent education in the classics, philosophy and mathematics. However, his father forbade him from studying medicine, so Cardano trained secretly as a physician.

After some years wandering as an unlicensed doctor, Cardano became a professor of mathematics in Milan in 1534. This led to his first mathematical writings on the theories of proportions and irrational numbers. Cardano also developed a reputation as a physician with innovative approaches to treating illnesses.

Cardano achieved major fame across Europe with the publication of his book Ars Magna in 1545. It contains the first published solutions to cubic and quartic equations and is widely considered one of the greatest scientific discoveries of its time.

In "Ars Magna", Cardano provided a method to solve cubic equations ($Ax^3 + bx^2 + cx + d = 0$), a problem that had puzzled mathematicians for centuries. The solution, now known as "Cardano's Method," involves a clever transformation of the original equation into a simpler one, using the symmetries of the cube. It was a groundbreaking achievement that expanded the horizon of algebraic solutions.

It's important to note that Cardano's solution to the cubic equation had actually been discovered by the mathematician Scipione del Ferro, and later rediscovered by Cardano's student Lodovico Ferrari. Cardano, learning of del Ferro's secret solution after his death, decided to publish the method, crediting del Ferro.

Cardano's other great mathematical contribution came in his book Liber de Ludo Aleae published in

1663 which covered mathematics of dice games and gambling. In it, he calculated the first probabilistic odds and introduced vital concepts like sample spaces, combinations and counting principles. This early study of probability greatly influenced later mathematicians like Fermat and Pascal.

Cardano was also one of the pioneers in the systematic study of probability. His significant contributions laid the groundwork for the theory of probability as we understand it today. His love of gambling prompted him to look for mathematical principles in games of chance, which led to his groundbreaking work, "Liber de Ludo Aleae" or "Book on Games of Chance", written in the 1560s but published posthumously in 1663.

In "Liber de Ludo Aleae", Cardano was the first to provide a comprehensive analysis of gambling, including a deep mathematical and statistical analysis of various games, particularly games involving dice. He didn't merely rely on luck; he calculated probabilities to try and gain an edge in the games he played.

For instance, consider a simple game where two players roll a fair six-sided die, and the one with the higher number wins. What's the probability of winning such a game? Cardano reasoned it out like this: there are 36 possible outcomes (6 outcomes for the first player's roll, and 6 for the second, making 6x6=36 in total). In 6 of these outcomes, the players would draw. That leaves 30 outcomes, and since the dice are fair, 15 of these would be wins for the first player and 15 would be wins for the second player. So, the probability of winning is 15/36, which simplifies to 5/12 - slightly less than half, due to the possibility of a draw.

He also calculated the probability of rolling different sums with dice, recognizing that some outcomes are equally likely and others less so based on the number of favorable combinations. For example, there is only one way to roll a sum of 12 with 2 dice but 6 ways to roll a sum of 7.

In poker, he demonstrated how to calculate the probability of different hands by counting the combinations of cards that would produce pairs, three of a kind, full house etc. He showed a full house is rarer than a flush or straight.

In horse racing, he calculated odds by considering factors like the speed, endurance and prior performance of the horses or runners involved. These were primitive approximations of statistical reasoning. Similarly, he calculated life expectancy based on comparing death rates across ages and occupations. His rudimentary actuarial table allowed the calculation of annuities for the first time.

Beyond mathematical analysis, Cardano's book also touched on aspects of bluffing, managing one's money, and the psychology of gambling. He even provided advice on how to be a successful gambler, emphasizing the importance of deep knowledge of the game, understanding of the odds, a clear head, and a sizable bankroll. He warned against superstitions and betting systems, asserting that understanding the mathematical laws governing games was the key to success.

While Cardano did not create probability theory as we understand it today, his exploration of probability and his systematic approach to analyzing games of chance were revolutionary for the time. His ideas were a crucial stepping stone that other mathematicians built upon in the centuries that followed, leading to the development of modern probability theory and statistical science. Today, these fields have widespread applications, from insurance and finance to health sciences and artificial intelligence, standing as a testament to Cardano's pioneering work.

SRINIVASA RAMANUJAN

Srinivasa Ramanujan, an Indian mathematician, has a story that is as extraordinary as his mathematical genius. Born in 1887 in the town of Erode, in Tamil Nadu, India, Ramanujan's love for numbers was evident from a young age. This passion grew into an obsession, eventually leading him to make substantial contributions to mathematical analysis, number theory, and continued fractions.

Ramanujan grew up in a humble household. As a child, he independently mastered advanced mathematics through studying books borrowed from college students. He finished his primary education at Town Higher Secondary School, where he independently discovered advanced mathematical theorems and even developed his own theorems.

Ramanujan's early formal education was interrupted by poverty and illness. His brilliance was evident from an early age but his eccentricities like failing school exams made mentors skeptical. His obsession with mathematics led him to neglect other subjects, causing him to lose his scholarship to Government Arts College. Despite these setbacks, Ramanujan continued his mathematical work, recording his theorems and results in notebooks.

His breakthrough came when his work was recognized by G. H. Hardy, a British mathematician at the University of Cambridge. In his late teens, Ramanujan sent samples of his mathematical work to professors at the University of Madras. Impressed by the originality of his theorems, they became his patrons and connected him with G.H. Hardy in 1913. In a famous anecdote, Hardy rated the writings sent by Ramanujan in terms of 'quality of work' on a scale from 0 to 100. Hardy remarked that if his own work was worth 30, he would give Ramanujan's 100. Hardy then invited Ramanujan to Cambridge to collaborate, which launched his meteoric rise in the mathematics world.

In 1914, already in Cambridge, Ramanujan's talent fully bloomed. Despite having no formal training in pure mathematics, Ramanujan made substantial contributions to various mathematical fields. His work was characterized by a distinctive style of intuition, imagination, and insight that was seldom seen in the mathematical community.

His collaboration with Hardy was very fruitful but plagued by Ramanujan's eccentric habits and declining health. Ramanujan had an uncanny intuition that allowed him to conjecture deep results and compile about 4,000 theorems from his subconscious. He claimed goddess Namagiri would appear in dreams and present mathematical formulas for him to verify upon waking.

Ramanujan's most significant contributions were in the field of number theory, including properties of partitions, series, and continued fractions. One of his remarkable achievements was the discovery of an analytical expression for the number of partitions of a number 'n'. The function P(n) gives the number of distinct ways a given positive integer can be represented as a sum of other positive integers. For instance, the number 4 can be partitioned in 5 ways: 4, 3+1, 2+2, 2+1+1, and 1+1+1+1. So, P(4) equals 5. Ramanujan's formula significantly simplified the calculation of P(n) for large numbers 'n'.

n	#	Partitions	Partitions with only odd numbers
0	1	() empty partition	() empty partition
1	1	1	1
2	1	2	1+1
3	2	1+2, 3	1+1+1, 3
4	2	1+3, 4	1+1+1+1, 1+3
5	3	2+3, 1+4, 5	1+1+1+1+1, 1+1+3, 5

Ramanujan also explored the properties of numbers that could be expressed as the sum of two cubes in two different ways. According to the anecdote, Hardy once visited Ramanujan in a hospital in London and mentioned that the taxi he rode in had a very dull number, 1729. To this, Ramanujan immediately responded that 1729 was not dull at all as it was the smallest number that could be expressed as the sum of two cubes in two different ways: $1^3 + 12^3$ and $9^3 + 10^3$. This concept led to what is now famously known as the Hardy-Ramanujan number, or "taxicab number". A taxicab number is defined as an integer that can be expressed as a sum of n positive integer cubes in n distinct ways.

Number 2 is technically the smallest taxicab number as it can be expressed as $1^3 + 1^3$ or $1^3 + 1^3$. Therefore, 1,729, the number from Ramanujan's anecdote, would be the second smallest. The third and the fourth taxicab numbers are:

- Ta(3) = 87,539,319
  - $167^3 + 436^3$
  - $228^3 + 423^3$
  - $255^3 + 414^3$
- Ta(4) = 6,963,472,309,248
  - $2421^3 + 19083^3$
  - $5436^3 + 18948^3$
  - $10200^3 + 18072^3$
  - $13322^3 + 16630^3$

Despite his incredible intellect, Ramanujan's life was fraught with hardship and health issues. Sadly, Ramanujan's career was cut short when he fell severely ill in 1917. He returned to India in 1919 and died the following year at the age of 32. Doctors said his symptoms matched hepatic amoebiasis and vitamin deficiencies from his strict vegetarian diet.

Ramanujan left behind three dense notebooks and groundbreaking papers full of novel formulas, rife with possibilities. Mathematicians are still working to prove the intuitions found in his writings 100 years later.

For his thousands of original theorems, unorthodox creativity, and introduction of mathematical ideas that reshaped fields like analysis and number theory, Ramanujan earns admiration as one of history's most naturally gifted mathematicians. His story illustrates the inexhaustible mysteries of numbers and nature's infinite complexity.

PIERRE SIMON LAPLACE

In the annals of mathematical history, Pierre-Simon Laplace holds a place of high esteem. Born in 1749 in Normandy, France, Laplace rose from humble beginnings to become one of the most influential scientists of his time. His work in astronomy, statistics, and mathematical physics was revolutionary, earning him the nickname "The Newton of France."

Laplace's initial education was in theology, intended for a career in the church. However, his extraordinary aptitude for mathematics diverted his path. Encouraged by his mathematics teachers, Laplace moved to Paris at the age of 19 to pursue a career in the field. His intellect and prowess in mathematics quickly caught attention, and before long, he was appointed as a professor of mathematics at the École Militaire in Paris.

Laplace's brilliance shone particularly bright in the field of celestial mechanics. He carried forward the work of Isaac Newton, providing mathematical rigor and comprehensiveness to Newton's laws of motion and gravitation. Laplace published a five-volume masterpiece, "Mécanique Céleste" (Celestial Mechanics), which translated the geometry of Newton's "Principia" into the calculus that's used in modern physics.

One of Laplace's significant contributions in celestial mechanics was resolving a longstanding mystery related to the solar system. It was observed that the orbits of Jupiter and Saturn were irregular, which led to doubts about the stability of the solar system. Laplace developed an analytical theory explaining these irregularities as a resonance in the planets' orbits, asserting the stability of the solar system.

Laplace was also responsible for formulating the "Nebular Hypothesis," a theory about the formation of the solar system. He suggested that the solar system originated from a rotating mass of incandescent gas, which cooled and condensed into the sun and the planets. While his hypothesis was contested and refined over time, it was a pioneering attempt at explaining the formation of the solar system.

In the realm of probability theory and statistics, Laplace made several critical contributions. He helped to formalize Bayesian inference, a method of statistical reasoning that allows for the updating of hypotheses based on new evidence. This is embodied in what is known as "Laplace's Rule of Succession," which gives a rule for predicting the probability of a future event based on the occurrence of past events.

In addition, Laplace introduced the concept of a potential, which is fundamental in physics, particularly in electromagnetism and gravitation. He formulated "Laplace's equation," a second-order partial differential equation, used to describe a wide range of physical phenomena, such as the behavior of electric and gravitational potentials, fluid flow, and heat conduction.

Laplace's work also led to the development of the "Laplace Transform," a mathematical technique used to solve complex differential equations commonly found in engineering and physics. The Laplace Transform simplifies calculations by transforming the differential equations into simpler algebraic equations.

The French Revolution and the following Napoleonic era saw Laplace navigating the choppy waters of political change, serving various regimes while continuing his scientific work. He died in 1827, leaving behind a rich scientific legacy.

Laplace compiled and expanded upon previous work by giants like Newton and Euler to model the heavens using mathematics. His Mechanique Celeste represented the capstone of classical physics before the revolutions of electromagnetism and relativity. Laplace's probabilistic insights proved equally influential for the advancement of mathematical statistics. His singular contributions helped usher in the age of modern science, where precision measurement and mathematical law governed nature. For these profound achievements across astronomy, physics and mathematics, Laplace deserves recognition as one of the most accomplished intellects of the French Enlightenment.

BERNHARD RIEMANN

Georg Friedrich Bernhard Riemann, commonly known as Bernhard Riemann, was a German mathematician who made substantial contributions that shaped the landscape of modern mathematics. Born in 1826 in a small village in Lower Saxony, Germany, Riemann's profound influence on mathematics continues to resonate today, from geometry to number theory and complex analysis.

Riemann was born into a poor family, the second of six children. His father, a Lutheran pastor, educated Riemann until he went to secondary school. Despite a shy and introverted demeanor, Riemann demonstrated remarkable intellectual capabilities from a young age. After briefly studying theology and philology to follow his father's wishes, Riemann shifted to mathematics, which was his true passion.

In 1846, he went to the University of Göttingen, a prominent hub of mathematical research. Here, he was deeply influenced by Carl Friedrich Gauss, one of the leading mathematicians of the time. Riemann's doctoral dissertation, completed under Gauss's supervision, introduced the concept of Riemann surfaces, pioneering a new field of complex analysis known as Riemannian geometry.

In his work, Riemann replaced the Euclidean notion of flat surfaces with his new, more flexible "Riemann surfaces" which could twist and curve in multiple dimensions. He opened up the possibility of geometries where Euclid's parallel postulate does not hold – where parallel lines can intersect or diverge. This seminal work paved the way for Albert Einstein's General Theory of Relativity, as it provided a mathematical description of a universe with curved space.

After finishing his doctorate, Riemann went on to introduce the "Riemann integral," his way of defining the definite integrals, which is fundamental in calculus. He addressed a series of issues in the foundations of real analysis and developed tools that are essential for mathematical physics.

One of Riemann's most celebrated contributions came in number theory, embodied in the famous "Riemann Hypothesis." In 1859, Riemann turned his attention to the distribution of prime numbers. Prime numbers are the building blocks of arithmetic – they are the numbers that can only be divided evenly by 1 and themselves. Despite their simplicity, primes do not appear at regular intervals, and understanding their distribution is a central question in number theory.

Riemann made an audacious conjecture about the distribution of prime numbers, linking them to the complex zeros of a function now known as the Riemann Zeta function.

The Riemann zeta function is defined as the infinite series:

$$\zeta(s) = 1/1^s + 1/2^s + 1/3^s + 1/4^s + ...$$

If you substitute 's' with '2', the Zeta function takes on a very surprising value: $\pi^2/6$. What is π doing there!? This result is not obvious at all and was a famous problem known as the Basel problem, solved by Euler in the 18th century; if you substitute 's' with '3', the Zeta function equals an irrational number known as the Apéry's constant (1.2020569...). As 's' grows, the result gets smaller.

The Riemann Hypothesis, which is related to the zeros of the Riemann Zeta function, has implications for the distribution of prime numbers. If proven true, it would tell us a lot about how prime numbers are spread out along the number line.

Despite concerted efforts by numerous mathematicians, the Riemann Hypothesis remains unproven to this day and is one of the seven "Millennium Prize Problems" for which the Clay Mathematics Institute offers a $1,000,000 prize for a correct solution.

Despite his immense contributions, Riemann's career was relatively short due to his ill health. His life was marked by frequent bouts of depression and illness, possibly tuberculosis, which he contracted during his studies. He died during a holiday in Italy in 1866, at the young age of 39, leaving a plethora of unfinished ideas.

Riemann's work established him as a visionary in mathematics. He introduced new concepts and tools that enabled unprecedented advancements in various mathematical fields. He had a profound effect on how we understand space, shaping the geometry of the universe itself. His influence permeates modern mathematics, from the smallest prime number to the expansive cosmos, forever engraving his name in the annals of mathematical history. His work has continued to inspire mathematicians more than a century after his death, testifying to the timeless brilliance of his ideas. The extraordinary ideas that Riemann proposed, many of which were far ahead of his time, continue to puzzle, inspire, and provide fertile ground for mathematical discovery.

PIERRE DE FERMAT

Pierre de Fermat, one of the great personalities in the world of mathematics, led a life that was as intriguing as the theorems and postulates he proposed. Born in 1607, in Beaumont-de-Lomagne, France, Fermat was a lawyer by profession, but it was his work in mathematics, done mostly as a hobby, that immortalized his name in the annals of scientific history.

Fermat's mathematical journey was largely self-directed. His professional duties left him with little time for academia, and he was mostly detached from the mathematical community of his time. His contributions, however, were profound and far-reaching, spanning across number theory, probability theory, and geometry.

Fermat is most renowned for his work in number theory, particularly for "Fermat's Last Theorem." Fermat's most lasting legacy stems from a simple marginal note he wrote on the book "Arithmetica" by the ancient Greek mathematician Diophantus. The note stated: "I have a truly marvelous demonstration of this proposition which this margin is too narrow to contain." This simple statement ignited one of the most famous intellectual quests in the history of mathematics.

Despite claiming proof, Fermat never wrote the proof down. He enjoyed posing challenges to other mathematicians by proposing theorems without providing demonstrations.

Fermat's Last Theorem posits that there are no three positive integers a, b, and c that satisfy the equation $a^n + b^n = c^n$ for any integer value of n greater than 2. This conjecture stumped mathematicians for centuries and became a sort of "Holy Grail" in the field of mathematics. It wasn't until 1994, more than 350 years after Fermat's death, that British mathematician Sir Andrew Wiles finally provided a proof.

Fermat's contributions were not confined to number theory. He and Blaise Pascal are often credited as the founders of probability theory. The two corresponded extensively, discussing problems related to games of chance. The two initiated important concepts like the expected value, which laid the groundwork for understanding risk and randomness, shaping fields as diverse as insurance, finance, and computer science.

In the field of geometry, Fermat devised a principle that plays a fundamental role in the science of light, known as Fermat's Principle or the Principle of Least Time. This principle, central to the field of optics, asserts that light takes the path that requires the least time when it travels from one point to another.

Fermat's Method of Adequality was an early attempt at understanding the concept of a derivative, a fundamental notion in calculus. Although his method was not as rigorous as the limit processes developed later, it was significant in the evolution of mathematical analysis.

Fermat was an unusual character in the history of science. He seldom published his work, preferring instead to communicate his findings in letters to friends or as challenges to the mathematical community. He was notorious for announcing his results without proof, leaving a trail of enigmatic theorems in his wake. His unique approach frustrated many contemporaries, but it undeniably added to the intrigue and enduring fascination surrounding his contributions.

Fermat passed away in 1665, leaving behind a legacy that continues to resonate through the world of mathematics. His theorems have become cornerstones in multiple mathematical fields. Despite his posthumous reputation as an elusive figure, Fermat's genius is unquestionable. His amateur passion for mathematics birthed ideas that would inspire countless professional mathematicians, earning him the unofficial title of the "Prince of Amateurs". His work continues to illuminate the structure and patterns of the numerical world, confirming that the realm of mathematics is not solely the domain of professionals. Fermat's life and work remind us that curiosity and a desire to explore the unknown are the heart of mathematical discovery.

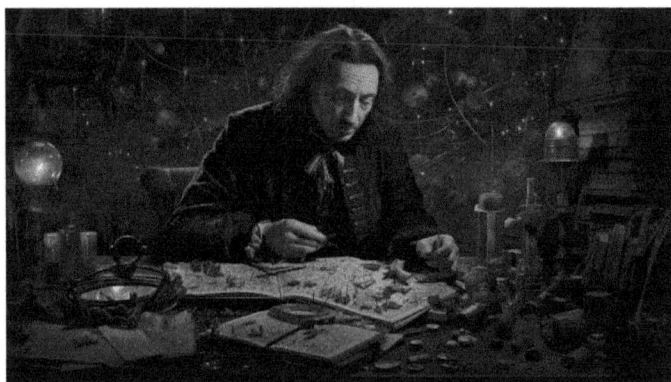

KATHERINE JOHNSON

Katherine Johnson, born as Creola Katherine Coleman in 1918, was an extraordinary mathematician whose calculations were integral to the success of numerous U.S. manned space missions. From a young age, her remarkable aptitude for mathematics was evident. Born in a time of racial segregation and limited opportunities for women, particularly in the sciences, Johnson's journey to NASA supporting the Mercury and Apollo missions, where she was a pivotal figure, is a remarkable story of perseverance, brilliance, and boundary-breaking.

Born and raised in White Sulphur Springs, West Virginia, Johnson's academic prowess was evident early on. She started high school at the age of 10 and graduated by 14, a stunning achievement given that education for African Americans, especially girls, was often interrupted or cut short in the early 20th century. She enrolled at West Virginia State, a

historically black college, at just 15 years old.
There, she soaked up as much knowledge as
possible, taking every math course available to her.

After graduating summa cum laude in 1937, Johnson started teaching at a black public school in Virginia. Then, a remarkable opportunity came in 1939 when West Virginia decided to quietly integrate its graduate schools. Katherine, alongside two men, were the first black students handpicked to integrate West Virginia University.

In 1953, she joined the National Advisory Committee for Aeronautics (NACA), the predecessor to NASA, in its guidance and navigation department. At NACA, she was initially consigned to a racially segregated computing unit. The group, known as the "West Area Computers," was comprised of African-American women whose role was to perform calculations for the engineers. The work was demanding, involving complex calculations done by hand or with mechanical calculators.

Despite the tedious work and the racially segregated workplace, Johnson and her colleagues persevered. Within two weeks of her hiring, Johnson was assigned to assist a project team, owing to her knowledge of analytic geometry. It was a temporary assignment, but her formidable skills became apparent, and she was soon permanently assigned to the branch.

Johnson's work at NACA was a mix of routine and pioneering. She crunched numbers and calculated trajectories, launch windows, and emergency return paths for many flights. She worked on the 1961 mission that made Alan Shepard the first American in space, the 1962 event that put John Glenn into orbit around Earth, and the 1969 Apollo 11 mission that sent humans to the moon.

One of her most notable contributions was during the Friendship 7 mission to orbit Earth, which astronaut John Glenn was to pilot. Glenn didn't trust the new electronic computing machines, which replaced the human computers, to calculate his trajectory correctly. He asked engineers to "get the girl"—referring to Johnson—to validate the calculations. Only after Johnson confirmed the computations did Glenn proceed with his mission, which turned out to be a great success.

She also co-authored a research report in 1960, "Determination of Azimuth Angle at Burnout for Placing a Satellite Over a Selected Earth Position," becoming the first woman in the Flight Research Division to receive credit as an author of a research report.

Johnson's mathematical genius and relentless spirit earned her a reputation for mastering complex manual calculations and helped her forge a path in a field heavily dominated by white men. Her work is an enduring legacy to the crucial role that African-American women played in the American space journey.

Johnson retired from NASA in 1986, after serving 33 years. However, it wasn't until 2015 that she was awarded the Presidential Medal of Freedom by President Barack Obama, the nation's highest civilian honor. Her life was depicted in the 2016 film "Hidden Figures," a portrayal that finally brought her the popular recognition she deserved.

Katherine Johnson passed away on February 24, 2020, leaving behind a legacy that continues to inspire people around the world. Her path-breaking journey as a black woman in the fields of mathematics and space science remains a beacon for future generations, demonstrating that with determination, brilliance, and courage, any barrier can be broken.

HENRI POINCARÉ

Born in 1854, in Nancy, France, Jules Henri Poincaré was a prodigious mathematician, theoretical physicist, and a philosopher of science. Renowned for his profound and wide-ranging contributions to mathematics and theoretical physics, Poincaré is often referred to as the last "Universalist" in math, as he excelled in all fields of the discipline as it existed during his lifetime.

From an early age, Poincaré demonstrated remarkable intellectual abilities. He completed his undergraduate studies at the prestigious École Polytechnique, where he showcased exceptional mathematical talent. After graduation, he continued his studies at the École des Mines. Later, he obtained his doctorate in mathematics from the University of Paris (Sorbonne) in 1879. He soon after became a lecturer at the University and began publishing influential papers. These caught the eye of Felix Klein, who helped arrange a professorship for Poincaré at the Sorbonne.

Poincaré's doctoral thesis was a major contribution to the field of differential equations, providing solutions to problems that had stumped other mathematicians. His work in this field led him to develop a new branch of mathematics called "qualitative theory of differential equations."

One of Poincaré's significant contributions to mathematics was in topology, sometimes referred to as 'rubber sheet geometry'. He introduced the concepts of homotopy and homology, which are fundamental to modern topology. His profound understanding of the geometric and topological properties of objects laid the foundation for algebraic topology.

Poincaré's Conjecture

Proposed in 1904, the Poincaré Conjecture is a central topic in the field of topology. Poincaré was investigating algebraic topology, which uses tools from abstract algebra to study topological spaces. He posed a seemingly simple question: If a three-dimensional space is closed (meaning it is finite and without any boundary), with no holes, is it equivalent to a three-dimensional sphere?

In other words, his conjecture proposed that if any loop in a three-dimensional space can be continuously tightened to a point without leaving the space, then the space is equivalent to a simple, ordinary sphere. It sounds straightforward, but it presented one of the most enduring problems in mathematics.

The Poincaré Conjecture was the first of the seven "Millennium Prize Problems" established by the Clay Mathematics Institute in 2000. These problems represent some of the most difficult unsolved problems in mathematics, with a prize of $1 million offered for the solution to each.

After nearly a century and numerous failed proofs, Poincaré's Conjecture was finally proven correct by the Russian mathematician Grigori Perelman in 2002-2003, who declined the award.

In number theory, Poincaré introduced an important generalization of continued fractions, known as the Poincaré series. His contributions to this field had a profound impact on the direction of mathematical research over the next century.

But perhaps Poincaré is best known for his work in celestial mechanics and the three-body problem. The latter is the challenge of predicting the motion of three bodies interacting with one another through gravitational pull, a problem that dates back to Newton. Poincaré demonstrated that there is no general analytical solution for the three-body problem, and instead, solutions can be approximated.

This work led him to lay the foundations of chaos theory, as he discovered that even with slight changes to the initial conditions of a system, the system could behave in significantly different ways over time. This idea is a key concept in what we now call 'chaotic systems'.

Poincaré's contributions to theoretical physics were also substantial. He played a significant role in developing the theory of special relativity, including the Lorentz transformations. Interestingly, while he and Einstein arrived at many of the same conclusions regarding relativity, they approached the subject from entirely different perspectives.

He also contributed to the understanding of electromagnetic waves and their propagation, providing solutions to problems that James Clerk Maxwell's equations posed. These solutions are now fundamental principles in the study of light and the transfer of energy in the form of waves.

Even beyond his technical work, Poincaré left his mark on philosophy, especially in the philosophy of science. He famously said that "science is built up with facts, as a house is with stones. But a collection of facts is no more a science than a heap of stones is a house."

Henri Poincaré passed away in 1912, but his rich legacy persists. Poincaré was renowned as an intuitive problem solver who worked across mathematics both in concrete and highly conceptual directions. The Poincaré conjecture, one of the famous unsolved problems, was finally proven in 2002 using many tools Poincaré pioneered.

Through his tireless investigation of complex functions, differential equations, celestial mechanics and the nature of mathematics itself, Poincaré ranks among the most influential mathematicians of the late 19th and early 20th centuries.

GOTTFRIED WILHELM LEIBNIZ

Gottfried Wilhelm Leibniz, born in 1646 in Leipzig, Germany, was an intellectual giant whose reach extended to several fields. He was a mathematician, philosopher, diplomat, lawyer, and even a mechanical engineer. He is widely recognized for his independently developed infinitesimal calculus, which he co-invented with Sir Isaac Newton, albeit from different perspectives.

Leibniz was born in Leipzig in 1646. He earned a doctorate in law from the University of Altdorf but also studied mathematics, history and philosophy extensively as a young man. His broad interests and analytic skills led to appointments as a diplomatic envoy, librarian, and historian across German principalities.

In 1672, Leibniz became the privy councilor and librarian for John Frederick, Duke of Hanover. This stable position allowed him to focus intensely on his writings in many fields over the next four decades until his death in 1716.

The most significant of these was his development of infinitesimal calculus around the same time as Isaac Newton. Despite this simultaneous development, Leibniz and Newton had quite different approaches. Leibniz focused on the notation and formalism of calculus, developing the integral and differential notation used universally today.

Infinitesimal calculus, more commonly known as just "calculus", is a branch of mathematics that deals with rates of change and the accumulation of quantities. It has been instrumental in the advancement of many fields of study, including physics, engineering, economics, computer science, statistics, biology, and more. Here are a few examples:

1. **Physics:** Calculus is the language of physics. It's used to describe the physical world—everything from the motion of planets and the behavior of fluids to the behavior of light and the forces on a bridge. Calculus enables physicists to create mathematical models of physical

phenomena and solve these models to gain insight into the world around us.

2. **Engineering:** Engineers use calculus for system design and control. For example, in electrical engineering, calculus is used to design and analyze circuits. In civil engineering, calculus is used to calculate stress and strain on materials and to understand the forces in structures like bridges or buildings.

3. **Economics:** Economists use calculus to model and predict systems. This might include predicting future sales, understanding how interest rates affect the economy, or optimizing production.

4. **Biology:** Calculus is used to model and understand population dynamics, enzyme kinetics, and the spread of diseases.

5. **Computer Science:** In computer graphics, calculus is used to determine how 3D surfaces change light to create the images we see on the screen. Calculus is also used in machine learning to train models and make predictions.

6. **Statistics:** Calculus is fundamental in statistics for determining probabilities and understanding distributions. For instance, it is used in regression analysis and in establishing statistical models like the normal distribution.

7. **Medicine:** Calculus is used to model and understand biological systems in medicine. For example, it can help model the human body's response to a drug over time.

8. **Astronomy:** Calculus helps astronomers measure distances, predict celestial events, understand the rotation and revolution of celestial bodies, and more.

From understanding how the universe works to designing the next generation of technology, calculus is a powerful tool for describing and exploring the world.

This independent development of calculus led to a bitter dispute, known as the calculus priority dispute, between Leibniz and Newton, as each claimed that the other had stolen his work. The argument persisted long after their deaths, with the mathematical community eventually accepting that both men developed calculus independently.

Leibniz also made significant contributions to mechanical engineering. He developed the concept of a calculating machine, a mechanical device that could perform all four basic arithmetic operations, laying the groundwork for the modern field of computer science. His machine, known as the Stepped Reckoner, could add, subtract, multiply, and divide by mechanized processes.

His philosophical contributions are equally significant. Leibniz's work ranged from topics like metaphysics, epistemology, logic, philosophy of religion, as well as ethics, law, and politics. His philosophical work "Monadology" is considered a masterpiece where he posits the idea of "monads" as the fundamental unit of reality, pre-determining all future actions and events.

Leibniz's vast array of interests and contributions resulted in a correspondingly vast written output. His collected works, which include numerous books, essays, and letters on a variety of topics, fill many volumes. Remarkably, he also found time to serve as a diplomat and political advisor, tasks that involved extensive travel and even more writing.

Leibniz died in 1716, but his work continues to have a significant impact today, particularly in mathematics and philosophy. His keen insight and relentless pursuit of knowledge left an indelible mark on the world, proving that intellectual curiosity and tenacity can lead to breakthroughs across numerous fields.

ADA LOVELACE

Ada Lovelace, born on December 10, 1815, in London, England, was a pioneer of computing science. She took part in writing the first published program and was a computing visionary, recognizing for the first time that computers could do much more than just calculations. Ada's work, though not fully recognized during her lifetime, has solidified her place as a fundamental contributor to computer science and programming.

Born Augusta Ada Byron, Ada was the only legitimate child of the famous poet Lord George Gordon Byron and his wife, Lady Anne Isabella Milbanke Byron. Her father's notorious lifestyle, coupled with a tumultuous marriage, led to her parents separating a month after Ada's birth. Consequently, Ada never met her father, who died in Greece when she was eight years old.

Lady Byron, wary of her daughter developing her father's unpredictable temperament, sought to immerse Ada in the study of mathematics and science. This was unconventional for women in the 19th century, who were not traditionally provided the same educational opportunities as men.

As a teenager, Ada's mathematical talents caught the attention of Mary Somerville, a renowned Scottish astronomer and mathematician. Somerville introduced Ada to Charles Babbage, a professor of Mathematics at Cambridge, in 1833. This meeting, which took place when Ada was just 17 years old, began a professional collaboration and friendship that lasted until Ada's death.

Babbage was working on plans for a machine he called the Analytical Engine, an ambitious precursor to the modern computer. Ada was captivated by the machine and began working with Babbage. Despite the project never being completed, it was a significant first step towards the modern field of computer science.

In 1835, Ada married William King, becoming Lady King. They had three children together. In 1838, her husband was made Earl of Lovelace, and Ada became the Countess of Lovelace, hence the name Ada Lovelace.

In 1843, Ada translated an Italian article about Babbage's Analytical Engine. However, Ada's contribution was much more than a mere translation. She added her own notes, which were three times longer than the original article, explaining how the machine would work. In these notes, Lovelace included a method for calculating a sequence of Bernoulli numbers with the Analytical Engine. This method is now recognized as the world's first published algorithm intended for implementation on a computer, earning Ada Lovelace the title of the world's first computer programmer.

Interestingly, Lovelace's notes also encompassed visionary ideas about the potential capabilities of the machine, including the manipulation of symbols and creation of music, concepts far ahead of her time.

Throughout her life, Lovelace was plagued by health issues. Her health further declined after a bout of cholera in 1851, leading to her death from uterine cancer in 1852 at the age of 36.

Ada Lovelace's contributions to the field of computer science were not fully recognized until over a century after her death. However, in the digital age, Lovelace's work has been widely acknowledged, and she is celebrated as a pioneer of computer programming.

Today, Ada Lovelace Day, observed in mid-October, celebrates the achievements of women in science, technology, engineering, and mathematics (STEM). Ada's life serves as an inspiration for women in STEM and a reminder of the potential lost when individuals are barred from pursuing their interests due to societal constraints.

Ada Lovelace's story, though tinged with elements of tragedy, shines through history as a tale of brilliance and intellectual prowess. It reveals a woman who, against the norms of her time, managed to etch her name permanently in the annals of science and technology. Yet, one is left wondering, had she lived longer or been born in a different era, how much more could this extraordinary woman have accomplished?

TERENCE TAO

Born in 1975 in Australia to parents of Chinese descent, Terence Tao displayed his exceptional mathematical abilities from a young age.

Tao was an exceptionally gifted child, demonstrating remarkable skills in mathematics at a young age. He was taking university-level mathematics classes when he was only 9 years old. Among the participants of the Johns Hopkins' Study of Exceptional Talent program, Tao is one of only three children who have ever scored 700 or more on the SAT math section at the age of eight, with his own score being an impressive 760. Julian Stanley, who served as the Director of the Study of Mathematically Precocious Youth, remarked that in all his years of extensive search, Tao's mathematical reasoning capability was unmatched.

At 13, he won a gold medal in the International Mathematical Olympiad. By 16, he earned his bachelor's and master's degrees, and by 21, he earned his Ph.D.

After obtaining his Ph.D. from Princeton University, he went on to join the faculty of the University of California, Los Angeles (UCLA). His mathematical work is wide-ranging, covering areas like harmonic analysis, partial differential equations, combinatorics, and representation theory.

Perhaps the best-known highlight of his career is his work on the Green-Tao theorem. In 2004, Dr. Tao, in collaboration with Ben Green, who is a mathematician at the University of Cambridge in England, cracked a problem associated with the Twin Prime Conjecture. They achieved this by examining progressions of prime numbers— sequences of numbers with equal spacing. An instance of this would be the numbers 3, 7, and 11, which form a progression of primes with a gap of 4. However, the subsequent number in this series, 15, is not prime. What Dr. Tao and Dr. Green established is that it is always feasible, within the boundless universe of integers, to locate a sequence of prime numbers with identical spacing and of any given length.

Tao has received many prestigious awards for his work, including the Fields Medal in 2006, often described as the "Nobel Prize of Mathematics", for "his contributions to partial differential equations, combinatorics, harmonic analysis and additive number theory". He was also awarded the Breakthrough Prize in Mathematics in 2015 which comes with a cash gift of $3 million. Other awards include the Princess of Asturias Award and the Grande Médaille.

Besides his profound mathematical contributions, Tao is well-known for his efforts to make mathematics more accessible. He maintains a blog where he shares open problems, lectures, and his views on various mathematical topics. He has also written several textbooks and lecture notes, making his insights and understanding of complex mathematical concepts available to students and colleagues alike.

Terence Tao's career serves as an example of how far mathematical curiosity can take one person. His work has expanded our understanding of mathematics, and his commitment to sharing knowledge continues to inspire future generations of mathematicians.

JOHN NASH

John Forbes Nash Jr. was born on June 13, 1928, in West Virginia, USA. He made innovative contributions to game theory, and his life story was depicted in the 2001 film A Beautiful Mind.

Nash was the son of an electrical engineer father and a schoolteacher mother who encouraged his intellectual development from an early age. At a young age, Nash displayed signs of genius, reportedly teaching himself to read when he was just a toddler.

By the time Nash reached high school, his passion for mathematics and science had become apparent. His intellectual capabilities quickly surpassed the material presented in school, so he began to study advanced mathematics on his own. He spent his summers reading books on mathematics and physics, developing a broad base of knowledge that would later prove instrumental in his academic pursuits.

Upon graduating from high school, Nash attended the Carnegie Institute of Technology (now Carnegie Mellon University) in Pittsburgh on a full scholarship. He originally intended to become an engineer like his father, but he soon discovered a greater affinity for mathematics. He eventually changed his major and graduated with both a bachelor's and master's degree in mathematics in 1948.

At 20, Nash began his Ph.D. at Princeton University, where he immersed himself in the world of mathematics. Princeton, at the time, was a haven for many of the world's leading mathematicians, including Albert Einstein and John von Neumann, offering Nash an inspiring intellectual environment.

Despite his relatively quiet demeanor, Nash stood out because of his unconventional thinking. He often spent hours in the mathematics lounge with his fellow students, scribbling equations and problems on the blackboard. During this time, he began to formulate ideas that would shape his career and the field of game theory forever.

In 1950, Nash completed his Ph.D. dissertation, titled "Non-Cooperative Games". The 28-page work was brief compared to most dissertations, but its impact was profound. In it, Nash introduced the concept of the "Nash equilibrium," which would go on to revolutionize the field of game theory.

In simple terms, a Nash equilibrium is a state of a system where no player can benefit from changing their strategy while the other players keep theirs unchanged. It is a stable state of a system involving competition or conflict, where everyone's strategy is optimal against the strategies of others. This concept found extensive applications in various fields, most notably economics, where it helped in understanding and predicting the behavior of markets and economic policy.

The **Prisoner's Dilemma** is a classic example of a situation in game theory where individuals, who would benefit from cooperating, have a strong incentive to act in their own self-interest. It's used to illustrate how the decisions of individuals can lead to worse outcomes for everyone involved.

The dilemma is usually presented as follows:

Two criminals are arrested, but police can't convict them on the principal charge without a confession. So, they separate the prisoners and offer each the same deal: If you confess and your partner doesn't, you'll get a light sentence (let's say 1 year), and your partner will get a heavy sentence (10 years). If neither of you confess, we can still convict you on a lesser charge; you'll both get a moderate sentence (2 years). But if you both confess, you both get a heavy sentence, albeit not as heavy as if only one of you confessed (say, 5 years each).

Here's the dilemma: each prisoner has a choice between only two options - to confess or not to confess. The outcome of their decision depends on what the other decides to do.

If we consider the best decision for both as a group, it's clear that they should both keep quiet and receive the moderate sentence of 2 years each. But from the individual's perspective, there's a strong temptation to confess. Consider the thought process: If my partner doesn't confess, I should definitely confess to get the lighter sentence of 1 year. If my partner does confess, I should still confess to avoid the worst sentence of 10 years. So, no matter what the other does, each prisoner's individual logic tells them to confess!

This scenario illustrates how individual decision-making can lead to worse outcomes when people don't or can't cooperate, even if cooperation would lead to a better result for everyone. The Nash equilibrium has real-world implications and can be seen in the over-exploitation of natural resources, where individuals acting in their own self-interest can lead to the depletion or destruction of a shared resource, to the detriment of all.

After obtaining his Ph.D., Nash accepted a position at the Massachusetts Institute of Technology (MIT) as a C.L.E. Moore Instructor in the mathematics department. His time at MIT proved to be fruitful both professionally and personally. He made significant contributions to the fields of differential geometry and partial differential equations. He also met his future wife, Alicia Lopez-Harrison de Larde, a physics student from El Salvador, at MIT. They married in 1957.

In the late 1950s, Nash's life took a turn. He started showing signs of mental illness. He became paranoid, claimed that aliens were communicating with him, and began obsessing over numbers, particularly prime numbers. In 1959, Nash was involuntarily admitted to a hospital, where he was diagnosed with paranoid schizophrenia. This marked the beginning of a decades-long struggle with mental illness, characterized by periods of intense paranoia and hospitalizations.

During his illness, Nash's career took a backseat. He resigned from his position at MIT, and his relationship with his wife Alicia strained, leading to their divorce in 1963. For several years, Nash lived a nomadic lifestyle, traveling in Europe and the United States, often showing up at his old haunts like Princeton, where he became a spectral figure, scribbling unintelligible equations on blackboards in empty rooms.

Yet, through all the struggles, Nash never completely left the world of mathematics. During his lucid periods, he continued to contribute to the field, although these contributions were often overlooked or ignored due to his mental illness.

The 1970s marked a turning point for Nash. His condition gradually improved, in what medical professionals term as "spontaneous remission". He returned to the academic community, and even began to lecture again. In 1978, he was awarded the John von Neumann Theory Prize for his discovery of non-cooperative equilibria, now called Nash equilibria.

In 1994, Nash's life took another significant turn when he was awarded the Nobel Prize in Economic Sciences for his work on game theory, cementing his place among the most influential thinkers of the 20th century. His acceptance of the prize marked a high point in his long and often tumultuous career. The award not only recognized his groundbreaking work but also symbolized a triumph over the adversity he had faced due to his mental illness.

Nash's personal life also saw positive changes. He rekindled his relationship with Alicia, who had supported him through many of his struggles. They remarried in 2001, renewing their commitment to each other.

Nash's story was brought to mainstream attention through the 1998 biography "A Beautiful Mind" by Sylvia Nasar. The book chronicled Nash's journey from mathematical prodigy to Nobel laureate while battling schizophrenia. The book was later adapted into an Academy Award-winning film of the same name, starring Russell Crowe as Nash. You should watch it!

Despite the dramatization of certain aspects of his life for the movie, it successfully depicted Nash's remarkable intellect and his struggle with mental illness, contributing significantly to public understanding and de-stigmatization of mental health issues.

John Nash passed away, along with his wife Alicia, in a car accident in New Jersey in 2015. His contributions to game theory and mathematics, his perseverance in the face of personal adversity, and his ultimate recognition form the basis of a powerful legacy that will continue to influence generations to come.

GEORGE CANTOR

Georg Cantor, born on March 3, 1845, in St. Petersburg, Russia, to a family of musicians, developed a passion for mathematics at a young age, despite his father's desire for him to become an engineer. His profound intuition for numbers and the infinite would lead him to alter the world's understanding of mathematics.

In 1862, Cantor moved to Germany to study at the University of Berlin, where he developed his early mathematical talent under the guidance of personalities such as Karl Weierstrass and Leopold Kronecker. However, this mentor-student relationship with Kronecker would later transform into a bitter professional rivalry, with Kronecker vehemently opposing Cantor's work on set theory and infinity.

After obtaining his doctorate in 1867, Cantor took a teaching position at the University of Halle, where he would spend the rest of his academic career. He married Vally Guttmann in 1874, and they had six children.

The critical shift in Cantor's career came when he began pondering the concept of infinity, a topic that had been largely ignored by mathematicians since the time of the Greeks. In the 1870s, Cantor proved that there were as many points in any given segment of a line, no matter how small, as there were in the whole infinite line itself. This concept, now known as "Cantor's first uncountability proof," was radical and met with harsh criticism.

Cantor went further into the abstract with his revolutionary "diagonal argument," proving that the set of real numbers is "more infinite" than the set of natural numbers.

When Cantor says that the set of real numbers is "more infinite" than the set of natural numbers, he is referring to these different sizes of infinity. Here's how he came to this conclusion:

1. **Countable and Uncountable Sets**: Cantor categorized sets into "countable" and "uncountable" sets. Countable sets are those that can be matched one-to-one with the set of natural numbers (which includes integers like 1, 2, 3, etc.). Uncountable sets cannot be matched in this way.

2. **Natural Numbers and Real Numbers**: The set of natural numbers is countable, while the set of real numbers (which includes numbers like 1.1, 1.01, 1.001, etc.) is uncountable. This is because between any two real numbers, no matter how close they are, you can always find another real number. This means there are more real numbers than natural numbers, even though both sets are infinite.

Cantor's ground-breaking "diagonal argument" provided a proof for this. He showed that even if you tried to list all real numbers between 0 and 1, you could always construct a new real number that wasn't on the list. This effectively proved that the set of real numbers is "more infinite" than the set of natural numbers.

Beyond the criticism from his colleagues, Cantor's ideas brought about severe personal challenges. The backlash and isolation he experienced likely contributed to his mental health problems. He suffered from several breakdowns and was repeatedly hospitalized. Despite these challenges, Cantor continued to work, publish, and correspond with other mathematicians.

One anecdote of interest is Cantor's correspondence with Catholic theologians. Cantor, a devout Christian, believed that his work on the infinite was a way of understanding the nature of God. He reached out to the Vatican to argue that his discoveries about the infinite were compatible with the Church's teachings. His letters reveal a man deeply committed to reconciling his faith with his scientific pursuits.

In the last years of his life, Cantor faced financial difficulties, as well as declining health. Despite the hardship, he found some late recognition when the Royal Society awarded him the Sylvester Medal in 1904. His work was gradually accepted and profoundly influenced the development of mathematics in the 20th century.

Cantor passed away on January 6, 1918, in a sanatorium in Halle. Despite facing professional isolation, personal struggles, and mental health problems, his courageous exploration of the infinite transformed mathematics.

In retrospect, the impact of Cantor's work is undeniable. Yet, throughout his life, the reception of his revolutionary ideas was ambiguous, a vivid illustration of the often-blurred line between madness and genius. Cantor spent his life dancing on that line, and perhaps that's where he found his infinity.

Javier Sanz

JOHN VON NEUMANN

John von Neumann's story is a symphony of mind, mathematics, and machines. A narrative that spirals through disciplines and decades, touching upon aspects that shape our very existence today. Born in Budapest, Hungary, in 1903, into a family of great affluence, von Neumann's intellect sparkled even in his early years.

A child prodigy, he could multiply and divide large numbers in his head by age six and was infatuated with logic and structure. His dinner conversations with his father about mathematics and economics laid the foundation for a future filled with intellectual rigor.

Educated in Europe's finest institutions, von Neumann's mathematical ability saw no bounds. His PhD in mathematics by the age of 23 was merely a prelude to a breathtaking intellectual journey that would traverse uncharted territories:

1. **Quantum Mechanics**: Von Neumann's work in quantum mechanics provided the mathematical rigor that the field desperately needed.

   - *Mathematical Foundation*: Von Neumann's treatise on the mathematical foundations of quantum mechanics laid out the axiomatic approach to quantum theory. He formulated the concept of a density matrix, which is vital in quantum statistical mechanics.

2. **Game Theory**: Venturing into the intersection of mathematics, economics, and psychology, von Neumann co-authored "Theory of Games and Economic Behavior" with economist Oskar Morgenstern.

   - Minimax Theorem: This theorem is the foundation of the field of game theory, stating that in a two-player, zero-sum game, there exists a strategy that allows both players to minimize their maximum losses.

   - Economic Applications: Game theory, as established by von Neumann, has found wide applications in economics, including in the study of oligopolies, auctions, and bargaining scenarios.

3. **Computer Science**: Perhaps most famous for his role in the development of the electronic computer, von Neumann's architecture is the backbone of virtually every computer today. He conceptualized the stored-program computer, where instructions and data reside in the same memory space, a radical idea that revolutionized computing. He also co-created the Monte Carlo Simulation, which is a statistical sampling technique, widely used today in fields like finance, physics, and engineering for numerical approximations in problems that at first seem to have a deterministic solution.

4. **Nuclear Research**: During World War II, von Neumann's polymathic abilities found application in the Manhattan Project. His expertise in shock waves and fluid dynamics proved critical in understanding nuclear explosions, a contribution that left an indelible mark on history.

5. **Weather Modeling & Operations Research**: Von Neumann's interest in meteorology led him to develop early models for weather prediction using computers, in particular the 30-ton supercomputer ENIAC (Electronic Numerical Integrator), which back in 1946 was the fastest computer in the world. His work in operations research during WWII also set a new standard for efficiency in logistics, supply chain management, and military planning.

6. **Automata Theory**: Von Neumann dreamed of machines that could copy themselves, like a robot building another robot. It's an idea that helps us think about both machines and life itself.

7. **Education & Advisory Roles**: Beyond his research, von Neumann's impact as a professor and an advisor to the U.S. government in matters of science and defense was profound.

Von Neumann's untimely death in 1957, due to cancer, marked the end of an era. An era marked by an insatiable thirst for knowledge, interdisciplinary explorations, and an intellectual depth that few could fathom.

The story of John von Neumann is not merely a collection of achievements; it's a glimpse into the mind of a genius who saw patterns where others saw chaos, who ventured into domains with fearless curiosity, and who shaped the very world we inhabit.

But behind the brilliance and the equations, who was John von Neumann, the man? Anecdotes of his rapid calculations, his boisterous laughter, and his strategic skill at games paint a picture, but the canvas remains incomplete. The inner workings of his mind, the nuances of his relationships, and the depths of his philosophical ponderings leave us with an enigma.

In the vast architecture of John von Neumann's contributions, might there be hidden chambers and secret passages still waiting to be explored? That question, like the man himself, enters the spheres of complexity and intrigue, where the answers may be as elusive and extraordinary as von Neumann's own mind.

JOSEPH-LOUIS LAGRANGE

Born in Turin in 1736, the child that would one day be known as Lagrange was initially destined to follow in his father's footsteps and study law. A fortuitous stumble upon a mathematical paper at the age of 17, however, diverted his path and ignited a flame that would burn with mathematical curiosity throughout his life.

Young Lagrange dove headfirst into the world of numbers and equations, and by the time he reached adulthood, his name was synonymous with groundbreaking mathematical insights. He forged new pathways in algebra, developing what we now know as Lagrange's Theorem, a concept that opened doors into group theory. But his thirst for mathematical discovery was unquenchable.

In the calculus of variations, he penned the Euler-Lagrange equation, intricately weaving paths, functions, and calculus into a harmonious relationship. This equation laid the foundation for physics and bridged the gap between numbers and the natural world.

Lagrange was not merely confined to abstract thoughts; he reached for the stars, quite literally. In celestial mechanics, his analysis of the three-body problem formed the bedrock of our understanding of the heavenly bodies. His "Mécanique analytique" is a work of art, presenting mechanics in a clear and elegant form, a trait he admired and always sought in his mathematical pursuits.

The **Three-Body Problem** is a captivating challenge in mathematics and physics. Imagine three celestial objects (three stars or planets) each exerting gravitational forces on the other two. They move in space, following the laws of gravity, and the problem is to predict exactly how they'll move over time.

If you have just two bodies, like the Earth and the Moon, you can predict exactly how they'll move around each other. The equations of motion are solvable, and you can calculate their paths precisely.

But if you just add one more body, everything changes. Suddenly, the equations become so complex that they defy a straightforward solution.

130

You can't write down a simple formula to describe the paths of all three bodies over time.

It's difficult because each body's motion is affected by the other two, and they all interact simultaneously. This creates a web of dependencies that makes it difficult to isolate one variable at a time and solve the equations in a tidy, analytical manner.

The Three-Body Problem isn't just a mathematical curiosity; it has real-world applications. Understanding the gravitational interactions between celestial bodies is vital for space missions, predicting the stability of planetary systems, and exploring fundamental questions about the universe.

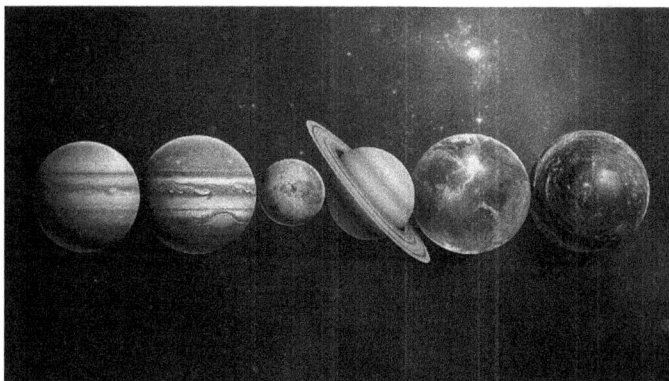

While we can't find a general, closed-form solution to the Three-Body Problem, it doesn't mean we're entirely in the dark. Numerical methods and computer simulations can approximate the behavior of the system, but they can't provide an exact, one-size-fits-all formula.

In number theory, Lagrange's footprints are deeply etched. He sought to untangle the knotted mysteries of Diophantine equations and laid the ground for the generations that would follow him.

But the life of Lagrange is not merely a tale of academic triumph. It is a story seasoned with peculiarities and rich anecdotes that render his biography as vibrant as one of his intricate mathematical proofs. He was known to be a modest genius, praising his contemporaries and often downplaying his own extraordinary accomplishments.

His physical health was fragile, yet his mind was anything but. A thinker to the core, Lagrange claimed that his best medicine was a good mathematical problem to solve. His obsession with mathematics was such that he often lost himself in thought, once boiling an egg for two hours and even misplacing a crucial paper, only to discover he'd used it to wrap butter.

Lagrange's influence knew no borders. From Turin to Berlin to Paris, he became a revered figure in academic circles, showered with honors and courted by prestigious institutions. Despite his weak constitution, his mind never waned, and he continued to contribute to mathematics up until his death in 1813.

The legacy of Lagrange is an awe-inspiring monument in the world of science. His ideas continue to resonate, shaping the way we see and understand the universe. But his life story is not without its shadows and complexities.

Lagrange often expressed dissatisfaction with his work, forever in pursuit of elusive simplicity. Did he view his monumental achievements as an unfulfilled mission? It's a question that seems to hang in the air, a riddle wrapped in the enigma that was Joseph-Louis Lagrange.

His story concludes, yet it doesn't feel complete. Like an unfinished theorem, the essence of Lagrange hovers between what we know and what we cannot grasp. In his brilliance and his quirks, in his triumphs and his unending search for mathematical purity, Lagrange remains both a beacon of intellectual achievement and a mysterious figure, forever calling us to explore, to understand, and to wonder at the beauty of a universe still unfolding.

Javier Sanz

Thank you!

Who was your favorite mathematician? My top 3 are Gauss, Pascal and Turing.

If this book helps you see the beauty in equations or just makes you chuckle, we'd be irrational with joy to hear about it! Toss us a review on Amazon. Tell us which part was your favorite, or which concept you're still scratching your head over (don't worry, we are too sometimes).

You can use this QR code to submit your review

Other Mathematicians like Sir Isaac Newton, Johann Bernoulli or James C. Maxwell were not featured in this book because I found their contributions to Physics were more notable. Find them on **Famous in STEM – Volume 1: The Greatest Physicists**

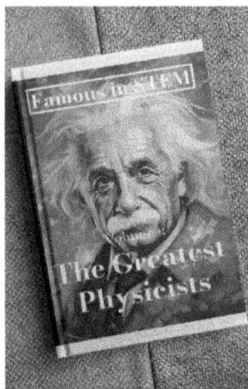

www.ingramcontent.com/pod-product-compliance
Lightning Source LLC
Chambersburg PA
CBHW060611200326
41521CB00007B/734